SUPERHERO SCIENCE

FLIGHT

BY BLAKE HOENA

BELLWETHER MEDIA • MINNEAPOLIS, MN

This edition first published in 2021 by Bellwether Media, Inc.

Library of Congress Cataloging-in-Publication Data

Names: Hoena, B. A., author.
Title: Flight / by Blake Hoena.
Description: Minneapolis : Bellwether Media, 2021. | Series: Blastoff! discovery : Superhero science | Includes bibliographical references and index. | Audience: Ages 7-13 | Audience: Grades 4-6 |
Summary: "Engaging images accompany information about the science of flight. The combination of high-interest subject matter and narrative text is intended for students in grades 3 through 8"– Provided by publisher.
Identifiers: LCCN 2020017763 (print) | LCCN 2020017764 (ebook) | ISBN 9781644872598 (library binding) | ISBN 9781681037226 (ebook)
Subjects: LCSH: Flight–Juvenile literature. | Aerodynamics–Juvenile literature. | Superheroes–Juvenile literature. | Transportation–Technological innovations–Juvenile literature.
Classification: LCC TL547 .H623 2021 (print) | LCC TL547 (ebook) | DDC 629.13–dc23
LC record available at https://lccn.loc.gov/2020017763
LC ebook record available at https://lccn.loc.gov/2020017764

Editor: Elizabeth Neuenfeldt   Designer: Jeffrey Kollock

Printed in the United States of America, North Mankato, MN.

TABLE OF CONTENTS

IRON MAN TAKES FLIGHT

Tony Stark is hard at work inside his lab. Suddenly FRIDAY, Tony's newest computer system, interrupts his thoughts. It notifies Tony of an attack downtown. It looks like a new villain wants to make a grand entrance.

Tony leaps to his feet, work forgotten. He taps his chest where there is a soft blue glow under his shirt. Red and gold metal pieces click together and cover Tony's entire body. Seconds later, he rushes outside and flies through the city to stop the attack. Tony Stark is Iron Man!

TONY STARK

BILLIONAIRE SUPERHERO

- Tony Stark is the head of the fictional tech company Stark Industries. But how rich is he? In 2019, people estimated Stark is worth up to $30 billion!

Iron Man's suit has boots that work like **miniature** jet engines. They help Iron Man zoom across the city! On his hands, special devices called repulsors help Iron Man easily dart around giant buildings.

IRON MAN ORIGINS

▸ **Iron Man first appeared in the *Tales of Suspense* comics in March 1963. He was created by writers Stan Lee and Larry Lieber and artists Don Heck and Jack Kirby.**

As Iron Man flies closer, he spots the villain surrounded by **debris**. Iron Man quietly flies toward the villain from above, making sure he is not seen. Then, Iron Man lands right behind the bad guy. Startled, the villain turns around. But before the villain can attack, Iron Man takes him down with a single repulsor blast. Iron Man's flying helped save the day!

SOARING SUPERHEROES

Many popular superheroes have the **ability** to fly. But they soar through the air in different ways. Superman was the first superhero who could fly. He was created by DC Comics in 1938. Superman is an alien from the planet Krypton. On Krypton, he did not have any special abilities. But Earth's yellow Sun gave him the power to fly!

In the 1940s, DC Comics introduced more flying heroes. Hawkman and Hawkgirl have wings made out of Nth metal. This material is found on their home planet of Thanagar. It resists **gravity**.

SUPERMAN

HAWKGIRL

SUPERHERO PROFILE

SUPERHERO NAME Captain Marvel

REAL NAME Carol Danvers

SUPERPOWERS

flight, energy blasts, super strength

BACKSTORY

Carol was an Air Force pilot when she was exposed to a Kree alien device called the Psyche-Magnitron. The device made her part human, part Kree. Over the years, she has fought alongside the Avengers and the X-Men. She is also known as Ms. Marvel and Binary.

SELECT APPEARANCES

- *Marvel Super-Heroes* comic number 13 (1968)
- *Ms. Marvel* comics (1977 to 1979)
- *Captain Marvel* comics (2014 to 2015)
- *Captain Marvel* feature film (2019)
- *Avengers: Endgame* feature film (2019)

As time passed, more heroes began to fly across the pages of comic books. In 1955, DC made Martian Manhunter and gave him the power of **telekinesis**. He uses this power to move himself and others up in the air!

Marvel Comics also made flying superheroes. In 1962, Marvel introduced Thor. He is a god who carries a magic hammer called Mjolnir. He can swing Mjolnir around to take flight! A few years later, Captain Marvel made her first appearance. She can zoom through space because she is part alien, part human!

Marvel continued to make flying heroes. In 1968, Vision was introduced. He has the power of **intangibility**. This allows him to become weightless and fly! Later, Falcon took to the skies with his EXO-7 suit. It has **mechanical** wings and a jetpack! In the 1970s, Storm started using her ability to control weather to fly. She uses the wind to **propel** herself through the air!

In recent decades, DC Comics created Miss Martian. She first appeared with her power of telekinesis in 2006. Like her mentor, Martian Manhunter, Miss Martian uses telekinesis to fly!

THE BIRD WHISPERER

▶ **In the comics, Falcon has a mental connection with birds. He even has a falcon named Redwing that helps him fight villains!**

MEDIA MANIA!

MOVIE *Aladdin*

YEARS 1992 and 2019

DESCRIPTION

The magic carpet from the movie is not a normal carpet. It can fly! Aladdin uses it to help him fly and defeat the evil sorcerer Jafar!

EVERYDAY FLIGHT

THE MONTGOLFIER BROTHERS IN A HOT-AIR BALLOON

Technology has helped people fly for centuries! In 1783, the Montgolfier brothers created an early hot-air balloon. Hot air inside a silk bag was lighter than the air outside. This helped the bag fly! The first powered flight occurred in 1903. Orville and Wilbur Wright attached a small engine to a glider. Their flight lasted 12 seconds and covered 120 feet (37 meters)!

Technology improved as time passed. In 1961, Yuri Gagarin became the first human to fly to space! By 1969, the first humans flew to the Moon. Today, people can travel anywhere on passenger planes.

THE WRIGHT FLYER

AIRPLANE OR BALLOON?

- **Blimps are a cross between airplanes and hot-air balloons. They have engines like airplanes and use gases to fly like a balloon!**

To take flight, all planes need **lift**. To produce lift, the wings on a plane are made so heavier air moves under the wings during flight. The heavy air below the wings lifts the plane into the air. To stay in the air, the lift has to be greater than the weight of the plane.

THE SCIENCE BEHIND FLIGHT

Planes also need **thrust** to fly. Thrust is created by powerful engines on a plane. To move forward, the amount of thrust needs to be greater than the amount of **drag**. Drag slows down planes as they fly.

A NINE-DAY TRIP

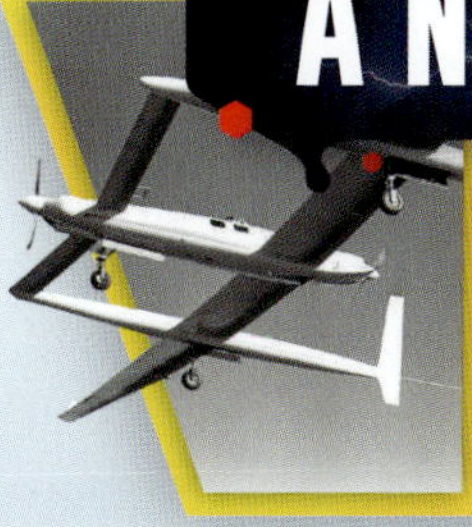

In 1986, the Rutan Voyager became the first plane to fly around the world nonstop! It took about 9 days to complete the flight.

IS HUMAN FLIGHT POSSIBLE?

Humans can pay for a plane ticket to fly across the world. But many people have wondered if humans can fly on their own like a bird.

Sadly, people cannot fly like a bird. Unlike people, birds have hollow bones. Their bodies are much lighter, which makes it easier for them to fly. Birds also have wide **wingspans** to support themselves while flying. For people to fly, they would need incredibly long wings to support their weight during flight.

INCREDIBLE WINGSPANS

- **Bald eagles stand about 3 feet (0.9 meters) tall. But their wingspans can be up to 7.5 feet (2.3 meters) long!**

FLIGHT IN NATURE

MANY ANIMALS AND INSECTS CAN TAKE TO THE SKY! HERE ARE SOME OF THE FASTEST FLYING ANIMALS.

PEREGRINE FALCON

CAN DIVE AT OVER 200 MILES PER HOUR | 322 KILOMETERS PER HOUR

BRAZILIAN FREE-TAILED BAT

CAN FLY AT UP TO 100 MILES PER HOUR | 161 KILOMETERS PER HOUR

DRAGONFLY

CAN FLY AT ABOUT 35 MILES PER HOUR | 56 KILOMETERS PER HOUR

Lacking the ability to fly has not stopped people from making technology to help them soar on their own. In 1919, Aleksandr Fyodorovich Andreyev **patented** the jetpack. A jetpack is a device that someone wears on their back. It has powerful jet engines so someone wearing it can fly.

THE ROCKETEER

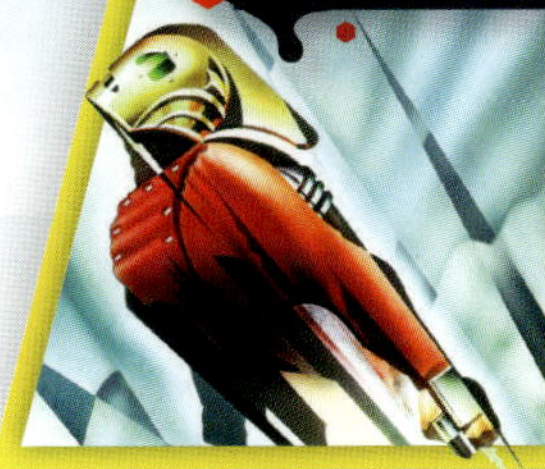

- Disney's 1991 movie *The Rocketeer* was about a superhero who uses a jetpack to save the day!

Since 1919, many people have tried to create their own working jetpacks. But jetpacks are not likely to be a common way to fly anytime soon. They use very expensive fuel. They can only provide short flights. They are also dangerous and difficult to control.

Today, people use wingsuits to try to glide through the air. Wingsuits have pieces of fabric stretched from a person's hips or feet to their arms to form wings. Another piece covers the area between their legs. To speed through the air, people jump from a high height and spread their arms and legs.

THE NEGATIVE EFFECTS OF FLIGHT ON HUMANS

COLD TEMPERATURES AT HIGH ALTITUDES

EFFECT: ▸ HYPOTHERMIA

THIN AIR AT HIGH ALTITUDES

EFFECT: ▸ DIFFICULTY BREATHING

But wingsuits can be deadly. People move at more than 100 miles (161 kilometers) per hour. They fly at high **altitudes** where air is colder and thinner. This can cause **hypothermia** and breathing problems.

THE FUTURE OF FLYING

Humans are not likely to develop superpowers and be able to fly like Superman. But people continue to seek new advancements in human flight. For example, British inventor Richard Browning designs jet engine-powered bodysuits.

Browning's invention turns the human body into an airplane. Jet engines are attached to each arm, and each leg has a wing. A person controls their flight with their body movements. In 2019, Browning flew at a record speed of 85 miles (137 kilometers) per hour in one of his suits.

JET ENGINE-POWERED BODYSUIT

JET ENGINE

Flying vehicles are another way people may take flight. In 2019, the company ALI built the XTURISMO. It is a hover bike. It hovers right above the ground with propellers. This will allow users to zoom around at high speeds!

HOVER BIKE

Drones could make human flight more common, too. People are working on making drones large enough to carry humans. These **remotely** flown aircraft could serve as flying taxis and avoid traffic on the ground!

FLYBOARD AIR
SHARK
ZAPATA

People could also take to the skies with the Flyboard Air. Flyboards have a longer flight time than jetpacks. Aboard them, people may be able to fly thousands of feet in the air. Flyboards might even reach speeds of 124 miles (200 kilometers) per hour.

People may not be able to fly on their own. But there continue to be new technologies that allow people to soar through the air. The sky is the limit!

GLOSSARY

ability—the power to do something

altitudes—heights above sea levels

debris—pieces left from something broken down or destroyed

drag—a force that slows down motion

drones—aircraft flown by computers or by people who are not on the aircraft

gravity—the force that pulls objects toward one another

hypothermia—a condition in which the body loses heat faster than it can produce it; hypothermia causes body systems to shut down.

intangibility—the power to not be made of physical substance

lift—an upward force that allows a plane to fly

mechanical—made or operated by a machine

miniature—very small

patented—gained the legal right to be the only person to make or sell an invention for a certain number of years

propel—to push

remotely—from a distance

telekinesis—the ability to move objects with one's mind

thrust—a forward or upward push

wingspans—the widths of bird wings; a wingspan is measured from the tip of one wing to the tip of the other.

TO LEARN MORE

AT THE LIBRARY

Enz, Tammy. *Soaring the Skies: Superman and the Science of Flight.* North Mankato, Minn.: Capstone Press, 2016.

Hutchison, Patricia. *The Science of Superheroes.* Minneapolis, Minn.: Abdo Publishing, 2017.

Kuromiya, Jun. *The Future of Transportation.* Minneapolis, Minn.: Lerner Publications, 2021.

ON THE WEB

FACTSURFER

Factsurfer.com gives you a safe, fun way to find more information.

1. Go to www.factsurfer.com.

2. Enter "flight" into the search box and click 🔍.

3. Select your book cover to see a list of related content.

INDEX

The images in this book are reproduced through the courtesy of: TCD/Prod.DB/ Alamy, front cover, pp. 3, 7 (Iron Man comics), 8 (Superman), 31; Photo 12/ Alamy, pp. 4-5 (Iron Man), 13 (Falcon); AF archive/ Alamy, pp. 5 (Tony Stark), 8-9, 10-11; Atlaspix/ Alamy, pp. 6-7; Allstar Picture Library/ Alamy, pp. 7, 13 (Aladdin); Album/ Alamy, p. 10 (Captain Marvel); iurii, pp. 12-13; DEA / G. NIMATALLAH/ Contributor/ Getty Images, pp. 14-15 (Montgolfier brothers); Science History Images/ Alamy, p. 15 (Wright Flyer); tonyz20, p. 15 (blimp); IM_photo, pp. 16-17; aapsky, p. 17 (fighter jet); Bettmann/ Contributor/ Getty Images, p. 17 (Rutan Voyager); JBKC, p. 18 (eagle); Edwin Butter, p. 19 (bald eagle); Harry Collins Photography, p. 19 (falcon); Dawn photos, p. 19 (bat); vnlit, p. 19 (dragonfly); Fairfax Media Archives/ Contributor/ Getty Images, pp. 20-21; PA Images/ Alamy, p. 21 (jetpack); sinesp, pp. 22-23; Guy Corbishley/ Alamy, pp. 24 (bodysuit), 24-25 (jet engine); Bloomberg/ Contributor/ Getty Images, pp. 26-27 (hover bike); ZUMA Press, Inc./ Alamy, pp. 26-27 (personal drone); Fit Ztudio, pp. 28-29; Andia/ Alamy, p. 29.